MY FIRST SCIENCE
BIOGRAPHY

Stephen Hawking

by Margaret Weber

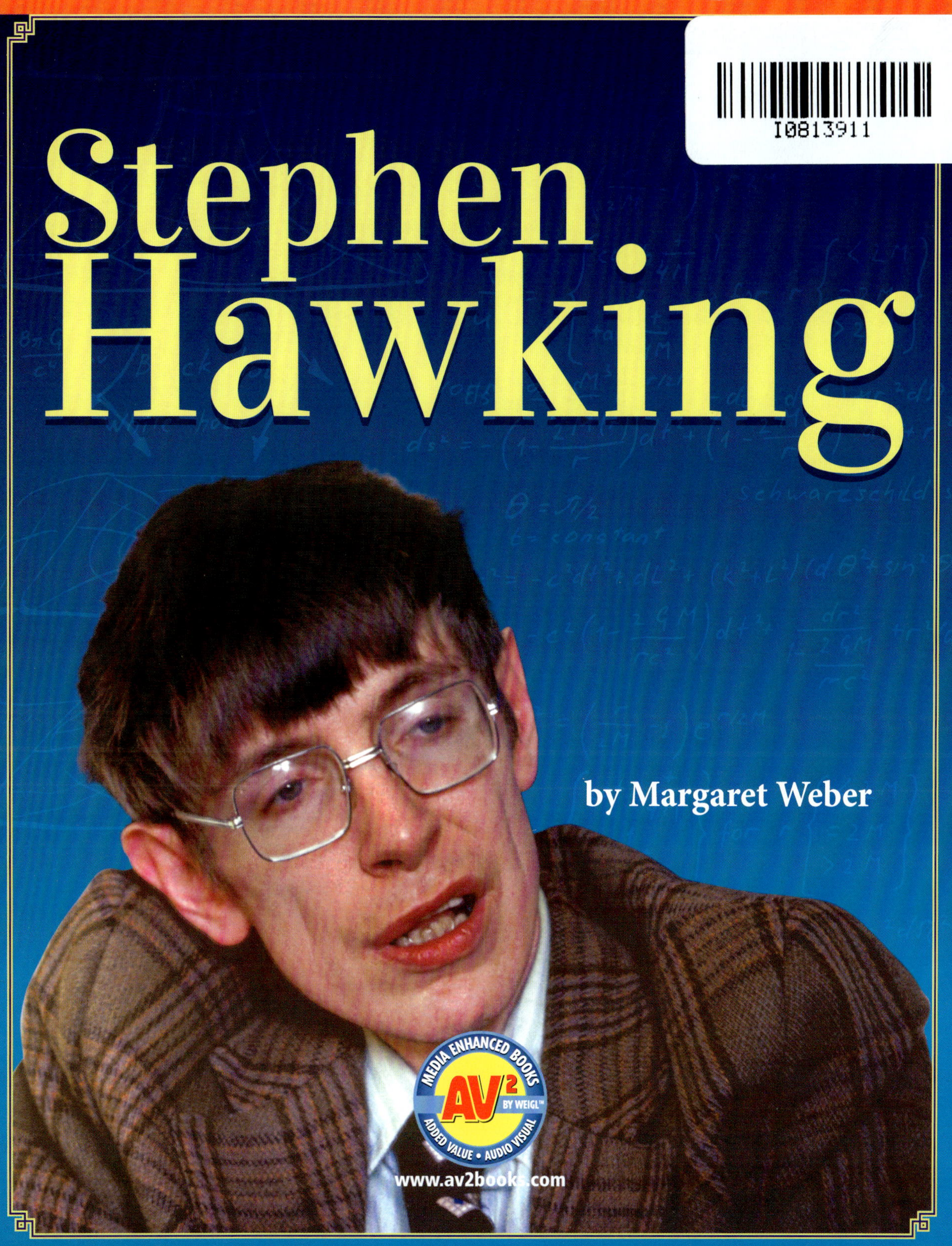

Go to www.av2books.com, and enter this book's unique code.

BOOK CODE

AVA33773

AV² by Weigl brings you media enhanced books that support active learning.

AV² provides enriched content that supplements and complements this book. Weigl's AV² books strive to create inspired learning and engage young minds in a total learning experience.

Your AV² Media Enhanced books come alive with...

Audio
Listen to sections of the book read aloud.

Video
Watch informative video clips.

Embedded Weblinks
Gain additional information for research.

Try This!
Complete activities and hands-on experiments.

Key Words
Study vocabulary, and complete a matching word activity.

Quizzes
Test your knowledge.

Slideshow
View images and captions, and prepare a presentation.

... and much, much more!

Published by AV² by Weigl
350 5th Avenue, 59th Floor
New York, NY 10118
Website: www.av2books.com

Library of Congress Control Number: 2019938596

ISBN 978-1-7911-0942-4 (hardcover)
ISBN 978-1-7911-0943-1 (softcover)
ISBN 978-1-7911-0944-8 (multi-user eBook)
ISBN 978-1-7911-0945-5 (single-user eBook)

Printed in Guangzhou, China
1 2 3 4 5 6 7 8 9 0 23 22 21 20 19

062019
311018

Project Coordinator: Heather Kissock
Art Director: Terry Paulhus

Photo Credits
Every reasonable effort has been made to trace ownership and to obtain permission to reprint copyright material. The publishers would be pleased to have any errors or omissions brought to their attention so that they may be corrected in subsequent printings.

Weigl acknowledges Getty, Alamy, iStock, and Shutterstock as its primary image suppliers for this title.

Stephen Hawking

Who Is Stephen Hawking?

Stephen Hawking is one of the best-known scientists of all time. Stephen studied how the **universe** began. He had many ideas about space.

Stephen had a disease called **ALS**. ALS affects the spinal cord and the brain. Stephen's illness did not stop his scientific work.

"Look up at the stars and not down at your feet. Try to make sense of what you see, and wonder about what makes the universe exist. Be curious."

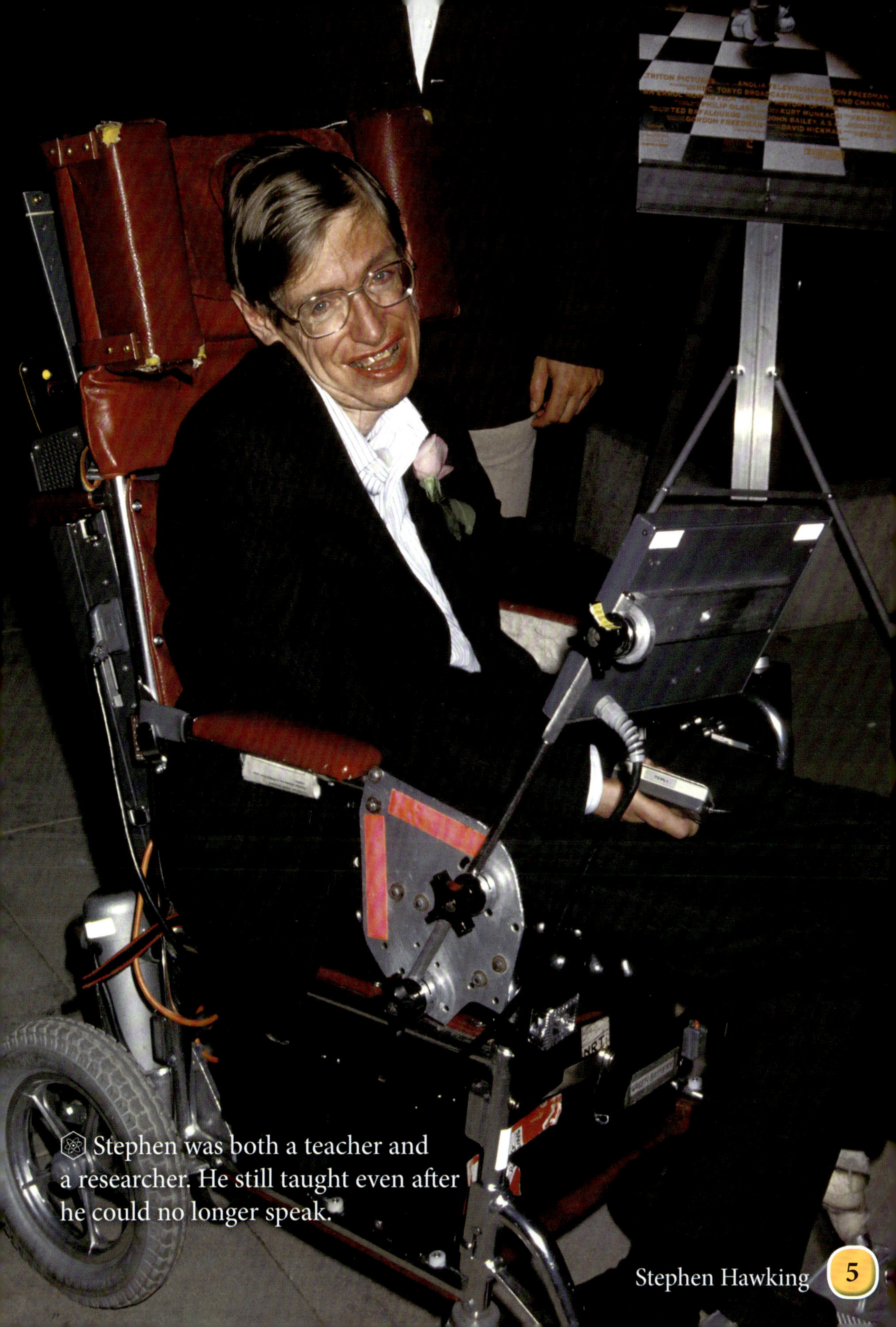

Stephen was both a teacher and a researcher. He still taught even after he could no longer speak.

Early Days

Stephen was born in 1942 in Oxford, England. His family moved to St Albans when he was 8 years old. Stephen studied at St Albans School. He showed a strong interest in science.

When he was older, Stephen went to the University of Oxford. He studied **physics** there. Stephen graduated when he was just 20 years old.

St Albans School is believed to have been founded in about 948 AD.

Where Is Oxford?

Oxford is a city in England. It is best known for its university. The University of Oxford is believed to be one of the best schools in the world.

Many scientists have studied at the University of Cambridge. Its library has more than 2 million books on its shelves.

Starting Out

Stephen wanted to continue his studies. He decided to go to the University of Cambridge. Stephen was interested in the stars and space. He wanted to understand how the universe formed. Stephen finished his program at the university in 1966.

Stephen began to write **theories** about space. He became very interested in **black holes**. Stephen studied black holes for many years.

In 1966, Stephen won the Adams Prize. The award is given to people doing important research in mathematics.

Influences

Stephen was influenced by many other scientists. He read their work to help him develop his own ideas. He also worked with other scientists. This helped him write his theories about the universe.

Stephen often talked about his work with his students and other scientists. Sharing ideas was important to him.

Roger Penrose

Stephen studied the work of Roger Penrose. Both men were interested in black holes.

Dennis Sciama

Dennis Sciama was a **mentor** to Stephen. He introduced Stephen to people who could help him with his work.

Albert Einstein

Albert Einstein was a well-known scientist. His ideas helped Stephen understand more about the rules of space.

Practice Makes Perfect

Stephen wanted to know how space and time were related. He studied space and pictures of space. He also read books and papers that other scientists had written. This is how Stephen came up with his biggest idea.

Stephen was able to show that the universe likely started with an explosion. This explosion is called the "Big Bang." After it happened, the universe began to form and expand.

Stephen wrote many books and essays to explain this idea. Other scientists loved his work. When he gave speeches, many people came to see him.

Stephen's book ***A Brief History of Time*** spent 147 weeks on the *New York Times* Best Sellers list.

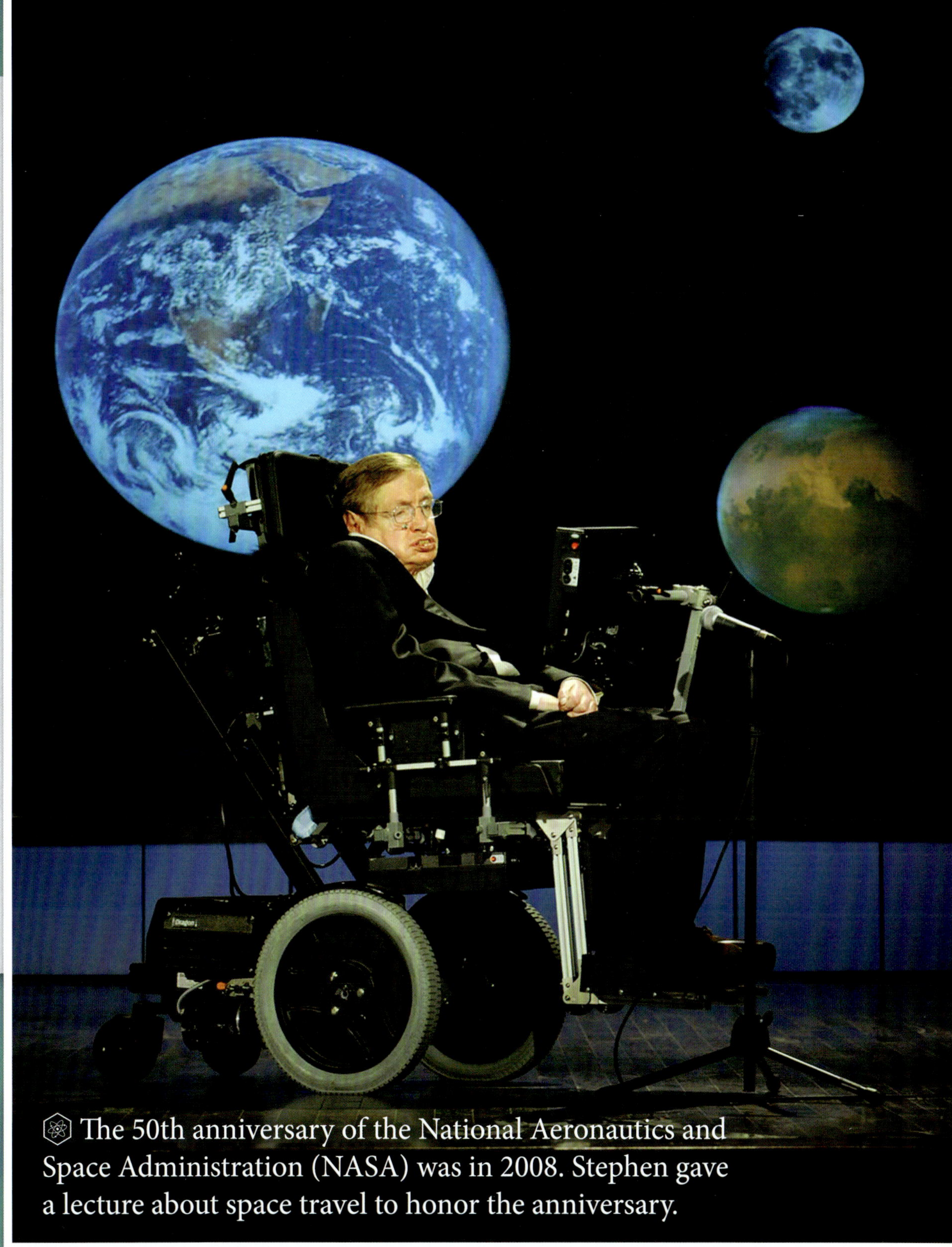

The 50th anniversary of the National Aeronautics and Space Administration (NASA) was in 2008. Stephen gave a lecture about space travel to honor the anniversary.

What Is a Scientist?

A scientist is someone who studies things. People who study space and time are called physicists. Scientists are very curious. They love solving problems. Scientists try to answer questions through **experiments**. They collect information using a six-step method.

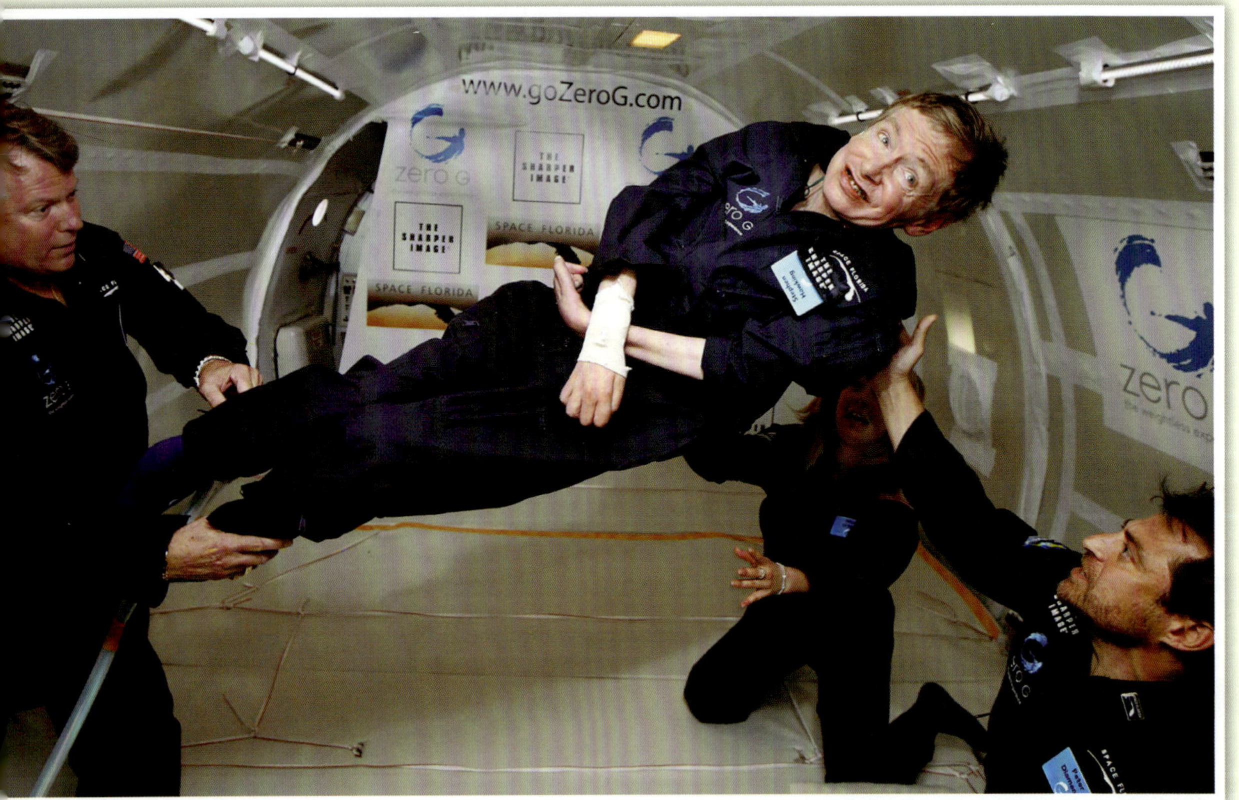

In 2007, Stephen rode in a special airplane to learn what it felt like to be in space.

Using a 6-Step Scientific Method

STEP 1

QUESTION

Scientists ask a question about what they want to learn. They read books and go online to research what other people know about the topic.

STEP 2

HYPOTHESIZE

The scientists then guess what the answer to their question might be. This guess is called a hypothesis.

STEP 3

EXPERIMENT

Scientists plan an experiment to see if their hypothesis is right. They gather the materials they need. Then, they set the materials up and do the experiment.

STEP 4

OBSERVE & RECORD

Scientists use their senses to **observe** what happens during their experiment. They watch. They smell. They listen, touch, and even taste. They then record what they find.

STEP 5

ANALYZE

Scientists think about what happened during their experiment. They decide if the experiment showed that the hypothesis was right or wrong.

STEP 6

SHARE RESULTS

Scientists let other people know about their experiment and what they found out. They may write a report or give a speech.

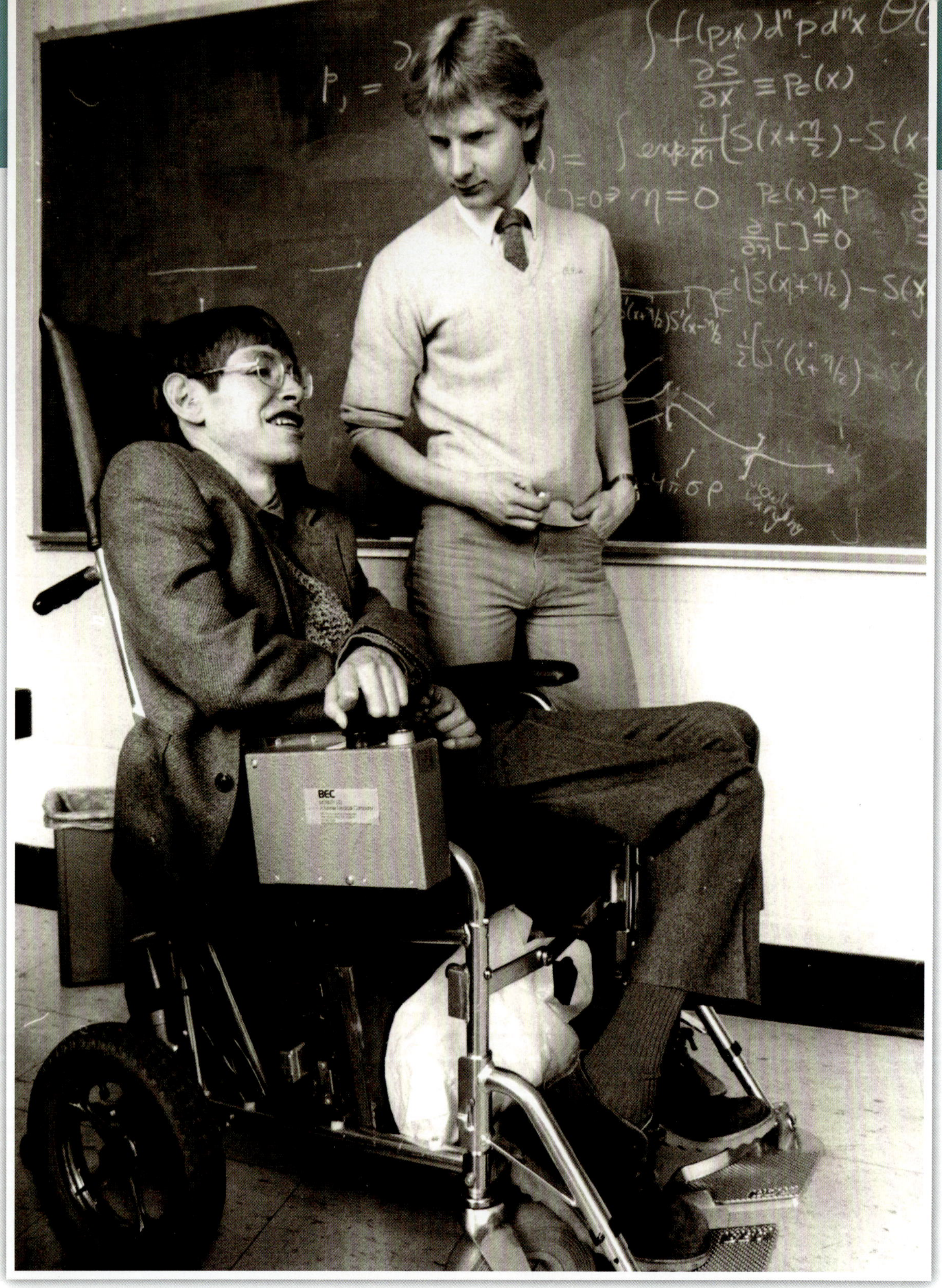

Stephen often had students and assistants help him with research and teaching.

Overcoming Obstacles

Stephen was 21 years old when he found out he had ALS. At first, he had trouble walking. Then, he had trouble speaking.

By 2008, Stephen could not use his hands or his voice. He worked with a computer company to make a special computer for him. This computer let him write and talk again. He used his cheek muscles to tell the computer what to do.

Stephen and the computer company were able to improve technology. Today, many people speak using computers.

Achievements and Successes

Stephen was given many awards during his life. In 2006, he won the Copley Medal. This is the one of the oldest and most important awards in science. It was given to Stephen for all of his scientific work. In 2009, he was awarded the Presidential Medal of Freedom. This award is given to people who have made a large impact on the world.

Stephen was also the author of 15 books. He even wrote five books for children. Stephen helped people of all ages understand space.

The Presidential Medal of Freedom was established by President John F. Kennedy in 1963.

The Copley Medal has been awarded to important scientists since 1731.

For Stephen's 70th birthday, the Science Museum in London had displays featuring different parts of Stephen's life.

Impact on Society

Stephen is remembered as an inspiration. Most people face obstacles in their lives. For those people, Stephen's life is seen as a great success story. He overcame many different challenges. Even those who are not scientists can admire him.

Today, physicists work to expand on Stephen's ideas. Scientists who look for black holes rely on his observations. His work gives scientists theories to test. This helps them understand what happens in space.

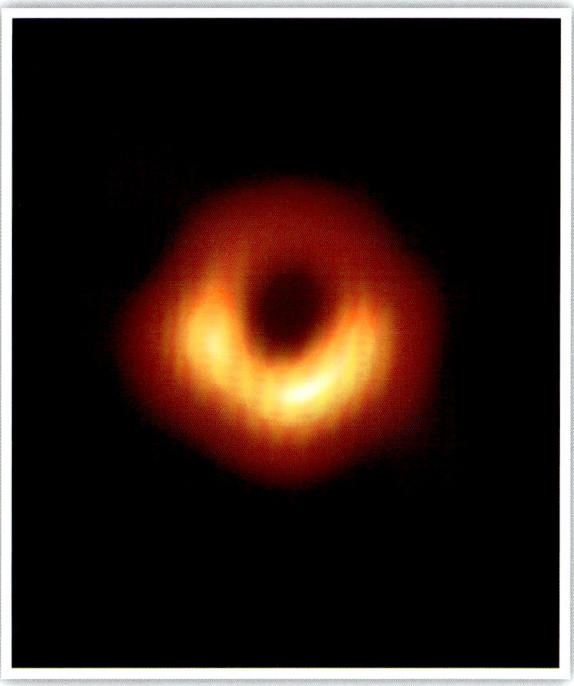

The first photo of a black hole was shown to the world on April 10, 2019.

Timeline

1942
Stephen is born on January 8 in Oxford.

1963
Stephen is told that he has ALS and that he has only two years to live.

1985
Stephen loses the ability to speak.

1988
Stephen's best-selling book, *A Brief History of Time*, is published.

2002
Stephen publishes his book, *The Theory of Everything*.

2018
Stephen dies on March 14 in Cambridge.

Key Words

ALS: amyotrophic lateral sclerosis, a disease that affects nerves in the brain and spinal cord

black holes: invisible objects in space whose gravity is so strong that even light cannot escape them

experiments: tests that are done to prove a hypothesis

mentor: someone who gives advice and helps someone else achieve his or her goals

observe: to carefully study an object or idea

physics: the study of energy, matter, and forces in the world

theories: ideas that try to explain a certain fact or event

universe: the space that contains all of the planets, stars, and galaxies

Index

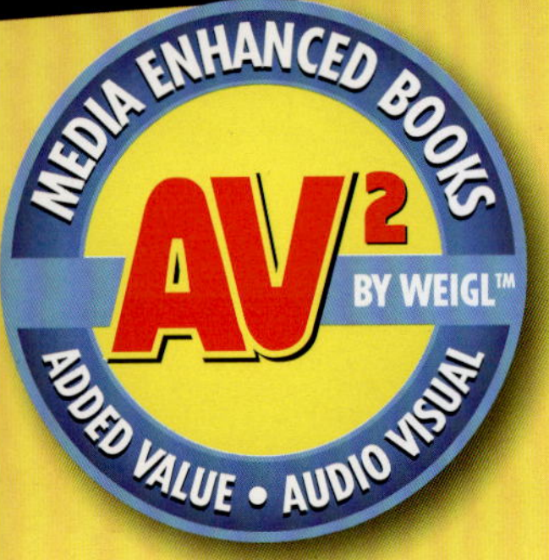

Log on to www.av2books.com

AV² by Weigl brings you media enhanced books that support active learning. Go to www.av2books.com, and enter the special code found on page 2 of this book. You will gain access to enriched and enhanced content that supplements and complements this book. Content includes video, audio, weblinks, quizzes, a slideshow, and activities.

AV² Online Navigation

Audio
Listen to sections of the book read aloud.

Book Pages
AV² pages directly correspond to pages in the book.

Video
Watch informative video clips.

Embedded Weblinks
Gain additional information for research.

Key Words
Study vocabulary, and complete a matching word activity.

Try This!
Complete activities and hands-on experiments.

Quizzes
Test your knowledge.

Slideshow
View images and captions, and prepare a presentation.

AV² was built to bridge the gap between print and digital. We encourage you to tell us what you like and what you want to see in the future.

Sign up to be an AV² Ambassador at www.av2books.com/ambassador.

Due to the dynamic nature of the internet, some of the URLs and activities provided as part of AV² by Weigl may have changed or ceased to exist. AV² by Weigl accepts no responsibility for any such changes. All media enhanced books are regularly monitored to update addresses and sites in a timely manner. Contact AV² by Weigl at 1-866-649-3445 or av2books@weigl.com with any questions, comments, or feedback.